THIS BOOK BELONGS TO

EMAIL:

ADDRESS:

CONTACT:

PHONE:

START DATE	END DATE

MO   TU   WE   TH   FR   SA   SU
☐   ☐   ☐   ☐   ☐   ☐   ☐

DATE: ___ / ___ / ___

PROJECT:

FOREMAN:

WEATHER

F° ___   C° ___   AM ___   PM ___

HOURS DUE TO BAD WEATHER	ISSUED AND DELAYS

NOTE:

COMPLETION DATE	DAYS AHEAD OF SCHEDULE	DAYS BEHIND SCHEDULE

SAFETY AND INCIDENTS

SAFETY ISSUES THAT NEED TO BE ADDRESSED	ACCIDENTS / INCIDENTS / STEPS NEEDED TO RESOLVE

SUMMARY OF THE WORK DONE TODAY

IMPORTANT NOTES

NAME	SIGNATURE

TODAY LABOR

INITIALS	TRADE	START	FINISH	PAID HOURS	OVERTIME	COMPANY
☐ EMPLOYEE ☐ CONTRUCTOR		AM	PM			
☐ EMPLOYEE ☐ CONTRUCTOR		AM	PM			
☐ EMPLOYEE ☐ CONTRUCTOR		AM	PM			
☐ EMPLOYEE ☐ CONTRUCTOR		AM	PM			
☐ EMPLOYEE ☐ CONTRUCTOR		AM	PM			
☐ EMPLOYEE ☐ CONTRUCTOR		AM	PM			
☐ EMPLOYEE ☐ CONTRUCTOR		AM	PM			
☐ EMPLOYEE ☐ CONTRUCTOR		AM	PM			

EQUIPMENT ON SITE	NO. OF UNITE	WORKING YES / NO

HIRED EQUIPMENT	NO. OF UNITE	EQUIPMENT RENTED	FROM	RATE

NAME: _____   SIGNATURE: _____

MO TU WE TH FR SA SU
☐   ☐   ☐   ☐   ☐   ☐   ☐   DATE: __/__/__

PROJECT: FOREMAN:

WEATHER ☁ ⛅ ☁ 🌨 ☀ 🌧 ⛈ | HOURS DUE TO | ISSUED AND DELAYS
         F°   C°   AM   PM | BAD WEATHER |

NOTE: _____

COMPLETION DATE	DAYS AHEAD OF SCHEDULE	DAYS BEHIND SCHEDULE

SAFETY AND INCIDENTS

SAFETY ISSUES THAT NEED TO BE ADDRESSED	ACCIDENTS / INCIDENTS / STEPS NEEDED TO RESOLVE

SUMMARY OF THE WORK DONE TODAY

IMPORTANT NOTES

NAME	SIGNATURE

TODAY LABOR

INITIALS	TRADE	START	FINISH	PAID HOURS	OVERTIME	COMPANY
☐ EMPLOYEE ☐ CONTRUCTOR		AM	PM			
☐ EMPLOYEE ☐ CONTRUCTOR		AM	PM			
☐ EMPLOYEE ☐ CONTRUCTOR		AM	PM			
☐ EMPLOYEE ☐ CONTRUCTOR		AM	PM			
☐ EMPLOYEE ☐ CONTRUCTOR		AM	PM			
☐ EMPLOYEE ☐ CONTRUCTOR		AM	PM			
☐ EMPLOYEE ☐ CONTRUCTOR		AM	PM			
☐ EMPLOYEE ☐ CONTRUCTOR		AM	PM			

EQUIPMENT ON SITE	NO. OF UNITE	WORKING YES / NO

HIRED EQUIPMENT	NO. OF UNITE	EQUIPMENT RENTED	FROM	RATE

NAME: SIGNATURE:

MO TU WE TH FR SA SU
☐   ☐   ☐   ☐   ☐   ☐   ☐

DATE: __ / __ / __

PROJECT:

FOREMAN:

WEATHER

F° ___   C° ___   AM ___   PM ___

HOURS DUE TO BAD WEATHER	ISSUED AND DELAYS

NOTE:

COMPLETION DATE	DAYS AHEAD OF SCHEDULE	DAYS BEHIND SCHEDULE

SAFETY AND INCIDENTS

SAFETY ISSUES THAT NEED TO BE ADDRESSED	ACCIDENTS / INCIDENTS / STEPS NEEDED TO RESOLVE

SUMMARY OF THE WORK DONE TODAY

IMPORTANT NOTES

NAME	SIGNATURE

TODAY LABOR

INITIALS	TRADE	START	FINISH	PAID HOURS	OVERTIME	COMPANY
☐ EMPLOYEE ☐ CONTRUCTOR		AM	PM			
☐ EMPLOYEE ☐ CONTRUCTOR		AM	PM			
☐ EMPLOYEE ☐ CONTRUCTOR		AM	PM			
☐ EMPLOYEE ☐ CONTRUCTOR		AM	PM			
☐ EMPLOYEE ☐ CONTRUCTOR		AM	PM			
☐ EMPLOYEE ☐ CONTRUCTOR		AM	PM			
☐ EMPLOYEE ☐ CONTRUCTOR		AM	PM			
☐ EMPLOYEE ☐ CONTRUCTOR		AM	PM			

EQUIPMENT ON SITE	NO. OF UNITE	WORKING YES / NO

HIRED EQUIPMENT	NO. OF UNITE	EQUIPMENT RENTED	FROM	RATE

NAME: SIGNATURE:

MO TU WE TH FR SA SU
☐   ☐   ☐   ☐   ☐   ☐   ☐

DATE: ___ / ___ / ___

PROJECT:

FOREMAN:

WEATHER

F°   C°   AM   PM

HOURS DUE TO BAD WEATHER	ISSUED AND DELAYS

NOTE:

COMPLETION DATE	DAYS AHEAD OF SCHEDULE	DAYS BEHIND SCHEDULE

SAFETY AND INCIDENTS

SAFETY ISSUES THAT NEED TO BE ADDRESSED	ACCIDENTS / INCIDENTS / STEPS NEEDED TO RESOLVE

SUMMARY OF THE WORK DONE TODAY

IMPORTANT NOTES

NAME	SIGNATURE

TODAY LABOR

INITIALS	TRADE	START	FINISH	PAID HOURS	OVERTIME	COMPANY
☐ EMPLOYEE ☐ CONTRUCTOR		AM	PM			
☐ EMPLOYEE ☐ CONTRUCTOR		AM	PM			
☐ EMPLOYEE ☐ CONTRUCTOR		AM	PM			
☐ EMPLOYEE ☐ CONTRUCTOR		AM	PM			
☐ EMPLOYEE ☐ CONTRUCTOR		AM	PM			
☐ EMPLOYEE ☐ CONTRUCTOR		AM	PM			
☐ EMPLOYEE ☐ CONTRUCTOR		AM	PM			
☐ EMPLOYEE ☐ CONTRUCTOR		AM	PM			

EQUIPMENT ON SITE	NO. OF UNITE	WORKING YES / NO

HIRED EQUIPMENT	NO. OF UNITE	EQUIPMENT RENTED	FROM	RATE

NAME: SIGNATURE:

MO TU WE TH FR SA SU
☐   ☐   ☐   ☐   ☐   ☐   ☐

DATE: /   /

PROJECT:

FOREMAN:

WEATHER

F°   C°   AM   PM

HOURS DUE TO
BAD WEATHER

ISSUED AND DELAYS

NOTE:

COMPLETION DATE	DAYS AHEAD OF SCHEDULE	DAYS BEHIND SCHEDULE

SAFETY AND INCIDENTS

SAFETY ISSUES THAT NEED TO BE ADDRESSED	ACCIDENTS / INCIDENTS / STEPS NEEDED TO RESOLVE

SUMMARY OF THE WORK DONE TODAY

IMPORTANT NOTES

NAME	SIGNATURE

TODAY LABOR

INITIALS	TRADE	START	FINISH	PAID HOURS	OVERTIME	COMPANY
☐ EMPLOYEE ☐ CONTRUCTOR		AM	PM			
☐ EMPLOYEE ☐ CONTRUCTOR		AM	PM			
☐ EMPLOYEE ☐ CONTRUCTOR		AM	PM			
☐ EMPLOYEE ☐ CONTRUCTOR		AM	PM			
☐ EMPLOYEE ☐ CONTRUCTOR		AM	PM			
☐ EMPLOYEE ☐ CONTRUCTOR		AM	PM			
☐ EMPLOYEE ☐ CONTRUCTOR		AM	PM			
☐ EMPLOYEE ☐ CONTRUCTOR		AM	PM			

EQUIPMENT ON SITE	NO. OF UNITE	WORKING YES / NO

HIRED EQUIPMENT	NO. OF UNITE	EQUIPMENT RENTED	FROM	RATE

NAME: SIGNATURE:

MO TU WE TH FR SA SU
☐   ☐   ☐   ☐   ☐   ☐   ☐

DATE: /   /

PROJECT:

FOREMAN:

WEATHER   F°   C°   AM   PM

HOURS DUE TO BAD WEATHER	ISSUED AND DELAYS

NOTE:

COMPLETION DATE	DAYS AHEAD OF SCHEDULE	DAYS BEHIND SCHEDULE

SAFETY AND INCIDENTS

SAFETY ISSUES THAT NEED TO BE ADDRESSED	ACCIDENTS / INCIDENTS / STEPS NEEDED TO RESOLVE

SUMMARY OF THE WORK DONE TODAY

IMPORTANT NOTES

NAME	SIGNATURE

TODAY LABOR

INITIALS	TRADE	START	FINISH	PAID HOURS	OVERTIME	COMPANY
☐ EMPLOYEE ☐ CONTRUCTOR		AM	PM			
☐ EMPLOYEE ☐ CONTRUCTOR		AM	PM			
☐ EMPLOYEE ☐ CONTRUCTOR		AM	PM			
☐ EMPLOYEE ☐ CONTRUCTOR		AM	PM			
☐ EMPLOYEE ☐ CONTRUCTOR		AM	PM			
☐ EMPLOYEE ☐ CONTRUCTOR		AM	PM			
☐ EMPLOYEE ☐ CONTRUCTOR		AM	PM			
☐ EMPLOYEE ☐ CONTRUCTOR		AM	PM			

EQUIPMENT ON SITE	NO. OF UNITE	WORKING YES / NO

HIRED EQUIPMENT	NO. OF UNITE	EQUIPMENT RENTED	FROM	RATE

NAME: SIGNATURE:

MO TU WE TH FR SA SU
☐   ☐   ☐   ☐   ☐   ☐   ☐

DATE: /   /

PROJECT:

FOREMAN:

WEATHER

F°____ C°____ ____ AM ____ PM

HOURS DUE TO
BAD WEATHER

ISSUED AND DELAYS

NOTE: _____

COMPLETION DATE	DAYS AHEAD OF SCHEDULE	DAYS BEHIND SCHEDULE

SAFETY AND INCIDENTS

SAFETY ISSUES THAT NEED TO BE ADDRESSED	ACCIDENTS / INCIDENTS / STEPS NEEDED TO RESOLVE

SUMMARY OF THE WORK DONE TODAY

IMPORTANT NOTES

NAME	SIGNATURE

TODAY LABOR

INITIALS	TRADE	START	FINISH	PAID HOURS	OVERTIME	COMPANY
☐ EMPLOYEE ☐ CONTRUCTOR		AM	PM			
☐ EMPLOYEE ☐ CONTRUCTOR		AM	PM			
☐ EMPLOYEE ☐ CONTRUCTOR		AM	PM			
☐ EMPLOYEE ☐ CONTRUCTOR		AM	PM			
☐ EMPLOYEE ☐ CONTRUCTOR		AM	PM			
☐ EMPLOYEE ☐ CONTRUCTOR		AM	PM			
☐ EMPLOYEE ☐ CONTRUCTOR		AM	PM			
☐ EMPLOYEE ☐ CONTRUCTOR		AM	PM			

EQUIPMENT ON SITE	NO. OF UNITE	WORKING YES / NO

HIRED EQUIPMENT	NO. OF UNITE	EQUIPMENT RENTED	FROM	RATE

NAME: SIGNATURE:

MO   TU   WE   TH   FR   SA   SU
☐   ☐   ☐   ☐   ☐   ☐   ☐

DATE: /   /

PROJECT:

FOREMAN:

WEATHER

F°   C°   AM   PM

HOURS DUE TO BAD WEATHER	ISSUED AND DELAYS

NOTE:

COMPLETION DATE	DAYS AHEAD OF SCHEDULE	DAYS BEHIND SCHEDULE

SAFETY AND INCIDENTS

SAFETY ISSUES THAT NEED TO BE ADDRESSED	ACCIDENTS / INCIDENTS / STEPS NEEDED TO RESOLVE

SUMMARY OF THE WORK DONE TODAY

IMPORTANT NOTES

NAME	SIGNATURE

TODAY LABOR

INITIALS	TRADE	START	FINISH	PAID HOURS	OVERTIME	COMPANY
☐ EMPLOYEE ☐ CONTRUCTOR			AM	PM		
☐ EMPLOYEE ☐ CONTRUCTOR			AM	PM		
☐ EMPLOYEE ☐ CONTRUCTOR			AM	PM		
☐ EMPLOYEE ☐ CONTRUCTOR			AM	PM		
☐ EMPLOYEE ☐ CONTRUCTOR			AM	PM		
☐ EMPLOYEE ☐ CONTRUCTOR			AM	PM		
☐ EMPLOYEE ☐ CONTRUCTOR			AM	PM		
☐ EMPLOYEE ☐ CONTRUCTOR			AM	PM		

EQUIPMENT ON SITE	NO. OF UNITE	WORKING YES / NO

HIRED EQUIPMENT	NO. OF UNITE	EQUIPMENT RENTED	FROM	RATE

NAME: SIGNATURE:

MO   TU   WE   TH   FR   SA   SU
☐   ☐   ☐   ☐   ☐   ☐   ☐

DATE: ___ / ___ / ___

PROJECT:

FOREMAN:

WEATHER

F° ___   C° ___   AM ___   PM ___

HOURS DUE TO BAD WEATHER	ISSUED AND DELAYS

NOTE:

COMPLETION DATE	DAYS AHEAD OF SCHEDULE	DAYS BEHIND SCHEDULE

SAFETY AND INCIDENTS

SAFETY ISSUES THAT NEED TO BE ADDRESSED	ACCIDENTS / INCIDENTS / STEPS NEEDED TO RESOLVE

SUMMARY OF THE WORK DONE TODAY

IMPORTANT NOTES

NAME	SIGNATURE

TODAY LABOR

INITIALS	TRADE	START	FINISH	PAID HOURS	OVERTIME	COMPANY
☐ EMPLOYEE ☐ CONTRUCTOR		AM	PM			
☐ EMPLOYEE ☐ CONTRUCTOR		AM	PM			
☐ EMPLOYEE ☐ CONTRUCTOR		AM	PM			
☐ EMPLOYEE ☐ CONTRUCTOR		AM	PM			
☐ EMPLOYEE ☐ CONTRUCTOR		AM	PM			
☐ EMPLOYEE ☐ CONTRUCTOR		AM	PM			
☐ EMPLOYEE ☐ CONTRUCTOR		AM	PM			
☐ EMPLOYEE ☐ CONTRUCTOR		AM	PM			

EQUIPMENT ON SITE	NO. OF UNITE	WORKING YES / NO

HIRED EQUIPMENT	NO. OF UNITE	EQUIPMENT RENTED	FROM	RATE

NAME: SIGNATURE:

MO TU WE TH FR SA SU
☐   ☐   ☐   ☐   ☐   ☐   ☐

DATE: ___ / ___ / ___

PROJECT:

FOREMAN:

WEATHER   F°____ C°____   ____ AM ____ PM

HOURS DUE TO BAD WEATHER	ISSUED AND DELAYS

NOTE: _____

COMPLETION DATE	DAYS AHEAD OF SCHEDULE	DAYS BEHIND SCHEDULE

SAFETY AND INCIDENTS

SAFETY ISSUES THAT NEED TO BE ADDRESSED	ACCIDENTS / INCIDENTS / STEPS NEEDED TO RESOLVE

SUMMARY OF THE WORK DONE TODAY

IMPORTANT NOTES

NAME	SIGNATURE

TODAY LABOR

INITIALS	TRADE	START	FINISH	PAID HOURS	OVERTIME	COMPANY
☐ EMPLOYEE ☐ CONTRUCTOR			AM	PM			
☐ EMPLOYEE ☐ CONTRUCTOR			AM	PM			
☐ EMPLOYEE ☐ CONTRUCTOR			AM	PM			
☐ EMPLOYEE ☐ CONTRUCTOR			AM	PM			
☐ EMPLOYEE ☐ CONTRUCTOR			AM	PM			
☐ EMPLOYEE ☐ CONTRUCTOR			AM	PM			
☐ EMPLOYEE ☐ CONTRUCTOR			AM	PM			
☐ EMPLOYEE ☐ CONTRUCTOR			AM	PM			

EQUIPMENT ON SITE	NO. OF UNITE	WORKING YES / NO

HIRED EQUIPMENT	NO. OF UNITE	EQUIPMENT RENTED	FROM	RATE

NAME: SIGNATURE:

MO TU WE TH FR SA SU
☐   ☐   ☐   ☐   ☐   ☐   ☐

DATE: __ / __ / __

PROJECT:

FOREMAN:

WEATHER

F°   C°   AM   PM

| HOURS DUE TO BAD WEATHER | ISSUED AND DELAYS |

NOTE: _____

COMPLETION DATE	DAYS AHEAD OF SCHEDULE	DAYS BEHIND SCHEDULE

SAFETY AND INCIDENTS

SAFETY ISSUES THAT NEED TO BE ADDRESSED	ACCIDENTS / INCIDENTS / STEPS NEEDED TO RESOLVE

SUMMARY OF THE WORK DONE TODAY

IMPORTANT NOTES

NAME	SIGNATURE

TODAY LABOR

INITIALS	TRADE	START	FINISH	PAID HOURS	OVERTIME	COMPANY
☐ EMPLOYEE ☐ CONTRUCTOR		AM	PM			
☐ EMPLOYEE ☐ CONTRUCTOR		AM	PM			
☐ EMPLOYEE ☐ CONTRUCTOR		AM	PM			
☐ EMPLOYEE ☐ CONTRUCTOR		AM	PM			
☐ EMPLOYEE ☐ CONTRUCTOR		AM	PM			
☐ EMPLOYEE ☐ CONTRUCTOR		AM	PM			
☐ EMPLOYEE ☐ CONTRUCTOR		AM	PM			
☐ EMPLOYEE ☐ CONTRUCTOR		AM	PM			

EQUIPMENT ON SITE	NO. OF UNITE	WORKING YES / NO

HIRED EQUIPMENT	NO. OF UNITE	EQUIPMENT RENTED	FROM	RATE

NAME: _____   SIGNATURE: _____

MO   TU   WE   TH   FR   SA   SU
☐   ☐   ☐   ☐   ☐   ☐   ☐

DATE: /   /

PROJECT:

FOREMAN:

WEATHER

F°____ C°____   ____ AM   ____ PM

HOURS DUE TO BAD WEATHER	ISSUED AND DELAYS

NOTE:

COMPLETION DATE	DAYS AHEAD OF SCHEDULE	DAYS BEHIND SCHEDULE

SAFETY AND INCIDENTS

SAFETY ISSUES THAT NEED TO BE ADDRESSED	ACCIDENTS / INCIDENTS / STEPS NEEDED TO RESOLVE

SUMMARY OF THE WORK DONE TODAY

IMPORTANT NOTES

NAME	SIGNATURE

TODAY LABOR

INITIALS	TRADE	START	FINISH	PAID HOURS	OVERTIME	COMPANY
☐ EMPLOYEE ☐ CONTRUCTOR		AM	PM			
☐ EMPLOYEE ☐ CONTRUCTOR		AM	PM			
☐ EMPLOYEE ☐ CONTRUCTOR		AM	PM			
☐ EMPLOYEE ☐ CONTRUCTOR		AM	PM			
☐ EMPLOYEE ☐ CONTRUCTOR		AM	PM			
☐ EMPLOYEE ☐ CONTRUCTOR		AM	PM			
☐ EMPLOYEE ☐ CONTRUCTOR		AM	PM			
☐ EMPLOYEE ☐ CONTRUCTOR		AM	PM			

EQUIPMENT ON SITE	NO. OF UNITE	WORKING YES / NO

HIRED EQUIPMENT	NO. OF UNITE	EQUIPMENT RENTED	FROM	RATE

NAME: SIGNATURE:

MO   TU   WE   TH   FR   SA   SU
☐   ☐   ☐   ☐   ☐   ☐   ☐

DATE: /   /

PROJECT:

FOREMAN:

WEATHER

F°   C°   AM   PM

HOURS DUE TO BAD WEATHER	ISSUED AND DELAYS

NOTE:

COMPLETION DATE	DAYS AHEAD OF SCHEDULE	DAYS BEHIND SCHEDULE

SAFETY AND INCIDENTS

SAFETY ISSUES THAT NEED TO BE ADDRESSED	ACCIDENTS / INCIDENTS / STEPS NEEDED TO RESOLVE

SUMMARY OF THE WORK DONE TODAY

IMPORTANT NOTES

NAME	SIGNATURE

TODAY LABOR

INITIALS	TRADE	START	FINISH	PAID HOURS	OVERTIME	COMPANY
☐ EMPLOYEE ☐ CONTRUCTOR		AM	PM			
☐ EMPLOYEE ☐ CONTRUCTOR		AM	PM			
☐ EMPLOYEE ☐ CONTRUCTOR		AM	PM			
☐ EMPLOYEE ☐ CONTRUCTOR		AM	PM			
☐ EMPLOYEE ☐ CONTRUCTOR		AM	PM			
☐ EMPLOYEE ☐ CONTRUCTOR		AM	PM			
☐ EMPLOYEE ☐ CONTRUCTOR		AM	PM			
☐ EMPLOYEE ☐ CONTRUCTOR		AM	PM			

EQUIPMENT ON SITE	NO. OF UNITE	WORKING YES / NO

HIRED EQUIPMENT	NO. OF UNITE	EQUIPMENT RENTED	FROM	RATE

NAME: SIGNATURE:

MO TU WE TH FR SA SU
☐   ☐   ☐   ☐   ☐   ☐   ☐

DATE: /   /

PROJECT:

FOREMAN:

WEATHER

F°   C°   AM   PM

HOURS DUE TO BAD WEATHER	ISSUED AND DELAYS

NOTE: _____

COMPLETION DATE	DAYS AHEAD OF SCHEDULE	DAYS BEHIND SCHEDULE

SAFETY AND INCIDENTS

SAFETY ISSUES THAT NEED TO BE ADDRESSED	ACCIDENTS / INCIDENTS / STEPS NEEDED TO RESOLVE

SUMMARY OF THE WORK DONE TODAY

IMPORTANT NOTES

NAME	SIGNATURE

TODAY LABOR

INITIALS	TRADE	START	FINISH	PAID HOURS	OVERTIME	COMPANY
☐ EMPLOYEE ☐ CONTRUCTOR		AM	PM			
☐ EMPLOYEE ☐ CONTRUCTOR		AM	PM			
☐ EMPLOYEE ☐ CONTRUCTOR		AM	PM			
☐ EMPLOYEE ☐ CONTRUCTOR		AM	PM			
☐ EMPLOYEE ☐ CONTRUCTOR		AM	PM			
☐ EMPLOYEE ☐ CONTRUCTOR		AM	PM			
☐ EMPLOYEE ☐ CONTRUCTOR		AM	PM			
☐ EMPLOYEE ☐ CONTRUCTOR		AM	PM			

EQUIPMENT ON SITE	NO. OF UNITE	WORKING YES / NO

HIRED EQUIPMENT	NO. OF UNITE	EQUIPMENT RENTED	FROM	RATE

NAME: SIGNATURE:

MO TU WE TH FR SA SU
☐   ☐   ☐   ☐   ☐   ☐   ☐

DATE: ___/___/___

PROJECT: _____

FOREMAN: _____

WEATHER

F°____ C°____   ____AM ____PM

| HOURS DUE TO BAD WEATHER | ISSUED AND DELAYS |

NOTE: _____

COMPLETION DATE	DAYS AHEAD OF SCHEDULE	DAYS BEHIND SCHEDULE

SAFETY AND INCIDENTS

SAFETY ISSUES THAT NEED TO BE ADDRESSED	ACCIDENTS / INCIDENTS / STEPS NEEDED TO RESOLVE

SUMMARY OF THE WORK DONE TODAY

IMPORTANT NOTES

NAME	SIGNATURE

TODAY LABOR

INITIALS	TRADE	START	FINISH	PAID HOURS	OVERTIME	COMPANY
☐ EMPLOYEE ☐ CONTRUCTOR		AM	PM			
☐ EMPLOYEE ☐ CONTRUCTOR		AM	PM			
☐ EMPLOYEE ☐ CONTRUCTOR		AM	PM			
☐ EMPLOYEE ☐ CONTRUCTOR		AM	PM			
☐ EMPLOYEE ☐ CONTRUCTOR		AM	PM			
☐ EMPLOYEE ☐ CONTRUCTOR		AM	PM			
☐ EMPLOYEE ☐ CONTRUCTOR		AM	PM			
☐ EMPLOYEE ☐ CONTRUCTOR		AM	PM			

EQUIPMENT ON SITE	NO. OF UNITE	WORKING YES / NO

HIRED EQUIPMENT	NO. OF UNITE	EQUIPMENT RENTED	FROM	RATE

NAME: SIGNATURE:

MO TU WE TH FR SA SU
☐   ☐   ☐   ☐   ☐   ☐   ☐

DATE: /   /

PROJECT:

FOREMAN:

WEATHER

F°____ C°____   AM____ PM____

HOURS DUE TO BAD WEATHER	ISSUED AND DELAYS

NOTE: _____

COMPLETION DATE	DAYS AHEAD OF SCHEDULE	DAYS BEHIND SCHEDULE

SAFETY AND INCIDENTS

SAFETY ISSUES THAT NEED TO BE ADDRESSED	ACCIDENTS / INCIDENTS / STEPS NEEDED TO RESOLVE

SUMMARY OF THE WORK DONE TODAY

IMPORTANT NOTES

NAME	SIGNATURE

TODAY LABOR

INITIALS	TRADE	START	FINISH	PAID HOURS	OVERTIME	COMPANY
☐ EMPLOYEE ☐ CONTRUCTOR		AM	PM			
☐ EMPLOYEE ☐ CONTRUCTOR		AM	PM			
☐ EMPLOYEE ☐ CONTRUCTOR		AM	PM			
☐ EMPLOYEE ☐ CONTRUCTOR		AM	PM			
☐ EMPLOYEE ☐ CONTRUCTOR		AM	PM			
☐ EMPLOYEE ☐ CONTRUCTOR		AM	PM			
☐ EMPLOYEE ☐ CONTRUCTOR		AM	PM			
☐ EMPLOYEE ☐ CONTRUCTOR		AM	PM			

EQUIPMENT ON SITE	NO. OF UNITE	WORKING YES / NO

HIRED EQUIPMENT	NO. OF UNITE	EQUIPMENT RENTED	FROM	RATE

NAME: SIGNATURE:

MO TU WE TH FR SA SU
☐   ☐   ☐   ☐   ☐   ☐   ☐

DATE: ___/___/___

PROJECT:

FOREMAN:

WEATHER

F° ___ C° ___   AM ___ PM ___

HOURS DUE TO BAD WEATHER	ISSUED AND DELAYS

NOTE: _____

COMPLETION DATE	DAYS AHEAD OF SCHEDULE	DAYS BEHIND SCHEDULE

SAFETY AND INCIDENTS

SAFETY ISSUES THAT NEED TO BE ADDRESSED	ACCIDENTS / INCIDENTS / STEPS NEEDED TO RESOLVE

SUMMARY OF THE WORK DONE TODAY

IMPORTANT NOTES

NAME	SIGNATURE

TODAY LABOR

INITIALS	TRADE	START	FINISH	PAID HOURS	OVERTIME	COMPANY
☐ EMPLOYEE ☐ CONTRUCTOR		AM	PM			
☐ EMPLOYEE ☐ CONTRUCTOR		AM	PM			
☐ EMPLOYEE ☐ CONTRUCTOR		AM	PM			
☐ EMPLOYEE ☐ CONTRUCTOR		AM	PM			
☐ EMPLOYEE ☐ CONTRUCTOR		AM	PM			
☐ EMPLOYEE ☐ CONTRUCTOR		AM	PM			
☐ EMPLOYEE ☐ CONTRUCTOR		AM	PM			
☐ EMPLOYEE ☐ CONTRUCTOR		AM	PM			

EQUIPMENT ON SITE	NO. OF UNITE	WORKING YES / NO

HIRED EQUIPMENT	NO. OF UNITE	EQUIPMENT RENTED	FROM	RATE

NAME: SIGNATURE:

MO   TU   WE   TH   FR   SA   SU
☐   ☐   ☐   ☐   ☐   ☐   ☐

DATE: /   /

PROJECT:

FOREMAN:

WEATHER

F°____ C°____   AM____ PM____

HOURS DUE TO BAD WEATHER	ISSUED AND DELAYS

NOTE: _____

COMPLETION DATE	DAYS AHEAD OF SCHEDULE	DAYS BEHIND SCHEDULE

SAFETY AND INCIDENTS

SAFETY ISSUES THAT NEED TO BE ADDRESSED	ACCIDENTS / INCIDENTS / STEPS NEEDED TO RESOLVE

SUMMARY OF THE WORK DONE TODAY

IMPORTANT NOTES

NAME	SIGNATURE

TODAY LABOR

INITIALS	TRADE	START	FINISH	PAID HOURS	OVERTIME	COMPANY
☐ EMPLOYEE ☐ CONTRUCTOR		AM	PM			
☐ EMPLOYEE ☐ CONTRUCTOR		AM	PM			
☐ EMPLOYEE ☐ CONTRUCTOR		AM	PM			
☐ EMPLOYEE ☐ CONTRUCTOR		AM	PM			
☐ EMPLOYEE ☐ CONTRUCTOR		AM	PM			
☐ EMPLOYEE ☐ CONTRUCTOR		AM	PM			
☐ EMPLOYEE ☐ CONTRUCTOR		AM	PM			
☐ EMPLOYEE ☐ CONTRUCTOR		AM	PM			

EQUIPMENT ON SITE	NO. OF UNITE	WORKING YES / NO

HIRED EQUIPMENT	NO. OF UNITE	EQUIPMENT RENTED	FROM	RATE

NAME:

SIGNATURE:

MO TU WE TH FR SA SU
☐   ☐   ☐   ☐   ☐   ☐   ☐   DATE: /   /

PROJECT: FOREMAN:

WEATHER   ☁ ⛅ ☁ 🌨 ☀ 🌧 ⛈   HOURS DUE TO   ISSUED AND DELAYS
                                          BAD WEATHER
           F°   C°   AM   PM

NOTE:

COMPLETION DATE	DAYS AHEAD OF SCHEDULE	DAYS BEHIND SCHEDULE

SAFETY AND INCIDENTS

SAFETY ISSUES THAT NEED TO BE ADDRESSED	ACCIDENTS / INCIDENTS / STEPS NEEDED TO RESOLVE

SUMMARY OF THE WORK DONE TODAY

IMPORTANT NOTES

NAME	SIGNATURE

TODAY LABOR

INITIALS	TRADE	START	FINISH	PAID HOURS	OVERTIME	COMPANY
☐ EMPLOYEE ☐ CONTRUCTOR		AM	PM			
☐ EMPLOYEE ☐ CONTRUCTOR		AM	PM			
☐ EMPLOYEE ☐ CONTRUCTOR		AM	PM			
☐ EMPLOYEE ☐ CONTRUCTOR		AM	PM			
☐ EMPLOYEE ☐ CONTRUCTOR		AM	PM			
☐ EMPLOYEE ☐ CONTRUCTOR		AM	PM			
☐ EMPLOYEE ☐ CONTRUCTOR		AM	PM			
☐ EMPLOYEE ☐ CONTRUCTOR		AM	PM			

EQUIPMENT ON SITE	NO. OF UNITE	WORKING YES / NO

HIRED EQUIPMENT	NO. OF UNITE	EQUIPMENT RENTED	FROM	RATE

NAME: _____   SIGNATURE: _____

MO TU WE TH FR SA SU
☐   ☐   ☐   ☐   ☐   ☐   ☐

DATE: ___/___/___

PROJECT:

FOREMAN:

WEATHER

F° C° ___ AM ___ PM

HOURS DUE TO BAD WEATHER	ISSUED AND DELAYS

NOTE: _____

COMPLETION DATE	DAYS AHEAD OF SCHEDULE	DAYS BEHIND SCHEDULE

SAFETY AND INCIDENTS

SAFETY ISSUES THAT NEED TO BE ADDRESSED	ACCIDENTS / INCIDENTS / STEPS NEEDED TO RESOLVE

SUMMARY OF THE WORK DONE TODAY

IMPORTANT NOTES

NAME	SIGNATURE

TODAY LABOR

INITIALS	TRADE	START	FINISH	PAID HOURS	OVERTIME	COMPANY
☐ EMPLOYEE ☐ CONTRUCTOR		AM	PM			
☐ EMPLOYEE ☐ CONTRUCTOR		AM	PM			
☐ EMPLOYEE ☐ CONTRUCTOR		AM	PM			
☐ EMPLOYEE ☐ CONTRUCTOR		AM	PM			
☐ EMPLOYEE ☐ CONTRUCTOR		AM	PM			
☐ EMPLOYEE ☐ CONTRUCTOR		AM	PM			
☐ EMPLOYEE ☐ CONTRUCTOR		AM	PM			
☐ EMPLOYEE ☐ CONTRUCTOR		AM	PM			

EQUIPMENT ON SITE	NO. OF UNITE	WORKING YES / NO

HIRED EQUIPMENT	NO. OF UNITE	EQUIPMENT RENTED	FROM	RATE

NAME: _____   SIGNATURE: _____

MO TU WE TH FR SA SU
☐   ☐   ☐   ☐   ☐   ☐   ☐

DATE: __/__/__

PROJECT:

FOREMAN:

WEATHER

F°___ C°___   AM___ PM___

HOURS DUE TO
BAD WEATHER

ISSUED AND DELAYS

NOTE: _____

COMPLETION DATE	DAYS AHEAD OF SCHEDULE	DAYS BEHIND SCHEDULE

SAFETY AND INCIDENTS

SAFETY ISSUES THAT NEED TO BE ADDRESSED	ACCIDENTS / INCIDENTS / STEPS NEEDED TO RESOLVE

SUMMARY OF THE WORK DONE TODAY

IMPORTANT NOTES

NAME	SIGNATURE

TODAY LABOR

INITIALS	TRADE	START	FINISH	PAID HOURS	OVERTIME	COMPANY
☐ EMPLOYEE ☐ CONTRUCTOR		AM	PM			
☐ EMPLOYEE ☐ CONTRUCTOR		AM	PM			
☐ EMPLOYEE ☐ CONTRUCTOR		AM	PM			
☐ EMPLOYEE ☐ CONTRUCTOR		AM	PM			
☐ EMPLOYEE ☐ CONTRUCTOR		AM	PM			
☐ EMPLOYEE ☐ CONTRUCTOR		AM	PM			
☐ EMPLOYEE ☐ CONTRUCTOR		AM	PM			
☐ EMPLOYEE ☐ CONTRUCTOR		AM	PM			

EQUIPMENT ON SITE	NO. OF UNITE	WORKING YES / NO

HIRED EQUIPMENT	NO. OF UNITE	EQUIPMENT RENTED	FROM	RATE

NAME: _____   SIGNATURE: _____

MO TU WE TH FR SA SU
☐   ☐   ☐   ☐   ☐   ☐   ☐

DATE: /   /

PROJECT:

FOREMAN:

WEATHER

F°____ C°____   AM____ PM____

HOURS DUE TO BAD WEATHER	ISSUED AND DELAYS

NOTE: _____

COMPLETION DATE	DAYS AHEAD OF SCHEDULE	DAYS BEHIND SCHEDULE

SAFETY AND INCIDENTS

SAFETY ISSUES THAT NEED TO BE ADDRESSED	ACCIDENTS / INCIDENTS / STEPS NEEDED TO RESOLVE

SUMMARY OF THE WORK DONE TODAY

IMPORTANT NOTES

NAME	SIGNATURE

TODAY LABOR

INITIALS	TRADE	START	FINISH	PAID HOURS	OVERTIME	COMPANY
☐ EMPLOYEE ☐ CONTRUCTOR		AM	PM			
☐ EMPLOYEE ☐ CONTRUCTOR		AM	PM			
☐ EMPLOYEE ☐ CONTRUCTOR		AM	PM			
☐ EMPLOYEE ☐ CONTRUCTOR		AM	PM			
☐ EMPLOYEE ☐ CONTRUCTOR		AM	PM			
☐ EMPLOYEE ☐ CONTRUCTOR		AM	PM			
☐ EMPLOYEE ☐ CONTRUCTOR		AM	PM			
☐ EMPLOYEE ☐ CONTRUCTOR		AM	PM			

EQUIPMENT ON SITE	NO. OF UNITE	WORKING YES / NO

HIRED EQUIPMENT	NO. OF UNITE	EQUIPMENT RENTED	FROM	RATE

NAME: SIGNATURE:

MO TU WE TH FR SA SU
☐   ☐   ☐   ☐   ☐   ☐   ☐

DATE: ___ / ___ / ___

PROJECT: _____

FOREMAN: _____

WEATHER

F°___ C°___ ___ AM ___ PM

HOURS DUE TO BAD WEATHER	ISSUED AND DELAYS

NOTE: _____

COMPLETION DATE	DAYS AHEAD OF SCHEDULE	DAYS BEHIND SCHEDULE

SAFETY AND INCIDENTS

SAFETY ISSUES THAT NEED TO BE ADDRESSED	ACCIDENTS / INCIDENTS / STEPS NEEDED TO RESOLVE

SUMMARY OF THE WORK DONE TODAY

IMPORTANT NOTES

NAME	SIGNATURE

TODAY LABOR

INITIALS	TRADE	START	FINISH	PAID HOURS	OVERTIME	COMPANY
☐ EMPLOYEE ☐ CONTRUCTOR		AM	PM			
☐ EMPLOYEE ☐ CONTRUCTOR		AM	PM			
☐ EMPLOYEE ☐ CONTRUCTOR		AM	PM			
☐ EMPLOYEE ☐ CONTRUCTOR		AM	PM			
☐ EMPLOYEE ☐ CONTRUCTOR		AM	PM			
☐ EMPLOYEE ☐ CONTRUCTOR		AM	PM			
☐ EMPLOYEE ☐ CONTRUCTOR		AM	PM			
☐ EMPLOYEE ☐ CONTRUCTOR		AM	PM			

EQUIPMENT ON SITE	NO. OF UNITE	WORKING YES / NO

HIRED EQUIPMENT	NO. OF UNITE	EQUIPMENT RENTED	FROM	RATE

NAME: SIGNATURE:

MO TU WE TH FR SA SU
☐   ☐   ☐   ☐   ☐   ☐   ☐

DATE: /   /

PROJECT:

FOREMAN:

WEATHER   [icons]

F°____ C°____   AM____ PM____

HOURS DUE TO
BAD WEATHER

ISSUED AND DELAYS

NOTE:

COMPLETION DATE	DAYS AHEAD OF SCHEDULE	DAYS BEHIND SCHEDULE

SAFETY AND INCIDENTS

SAFETY ISSUES THAT NEED TO BE ADDRESSED	ACCIDENTS / INCIDENTS / STEPS NEEDED TO RESOLVE

SUMMARY OF THE WORK DONE TODAY

IMPORTANT NOTES

NAME	SIGNATURE

TODAY LABOR

INITIALS	TRADE	START	FINISH	PAID HOURS	OVERTIME	COMPANY
☐ EMPLOYEE ☐ CONTRUCTOR		AM	PM			
☐ EMPLOYEE ☐ CONTRUCTOR		AM	PM			
☐ EMPLOYEE ☐ CONTRUCTOR		AM	PM			
☐ EMPLOYEE ☐ CONTRUCTOR		AM	PM			
☐ EMPLOYEE ☐ CONTRUCTOR		AM	PM			
☐ EMPLOYEE ☐ CONTRUCTOR		AM	PM			
☐ EMPLOYEE ☐ CONTRUCTOR		AM	PM			
☐ EMPLOYEE ☐ CONTRUCTOR		AM	PM			

EQUIPMENT ON SITE	NO. OF UNITE	WORKING YES / NO

HIRED EQUIPMENT	NO. OF UNITE	EQUIPMENT RENTED	FROM	RATE

NAME: SIGNATURE:

MO TU WE TH FR SA SU
☐   ☐   ☐   ☐   ☐   ☐   ☐

DATE: ___ / ___ / ___

PROJECT:

FOREMAN:

WEATHER

F°____ C°____   ____ AM ____ PM

| HOURS DUE TO BAD WEATHER | ISSUED AND DELAYS |

NOTE: _____

COMPLETION DATE	DAYS AHEAD OF SCHEDULE	DAYS BEHIND SCHEDULE

SAFETY AND INCIDENTS

SAFETY ISSUES THAT NEED TO BE ADDRESSED	ACCIDENTS / INCIDENTS / STEPS NEEDED TO RESOLVE

SUMMARY OF THE WORK DONE TODAY

IMPORTANT NOTES

NAME	SIGNATURE

TODAY LABOR

INITIALS	TRADE	START	FINISH	PAID HOURS	OVERTIME	COMPANY
☐ EMPLOYEE ☐ CONTRUCTOR		AM	PM			
☐ EMPLOYEE ☐ CONTRUCTOR		AM	PM			
☐ EMPLOYEE ☐ CONTRUCTOR		AM	PM			
☐ EMPLOYEE ☐ CONTRUCTOR		AM	PM			
☐ EMPLOYEE ☐ CONTRUCTOR		AM	PM			
☐ EMPLOYEE ☐ CONTRUCTOR		AM	PM			
☐ EMPLOYEE ☐ CONTRUCTOR		AM	PM			
☐ EMPLOYEE ☐ CONTRUCTOR		AM	PM			

EQUIPMENT ON SITE	NO. OF UNITE	WORKING YES / NO

HIRED EQUIPMENT	NO. OF UNITE	EQUIPMENT RENTED	FROM	RATE

NAME: SIGNATURE:

MO TU WE TH FR SA SU
☐   ☐   ☐   ☐   ☐   ☐   ☐

DATE: /   /

PROJECT:

FOREMAN:

WEATHER

F°   C°   AM   PM

HOURS DUE TO
BAD WEATHER

ISSUED AND DELAYS

NOTE:

COMPLETION DATE	DAYS AHEAD OF SCHEDULE	DAYS BEHIND SCHEDULE

SAFETY AND INCIDENTS

SAFETY ISSUES THAT NEED TO BE ADDRESSED	ACCIDENTS / INCIDENTS / STEPS NEEDED TO RESOLVE

SUMMARY OF THE WORK DONE TODAY

IMPORTANT NOTES

NAME	SIGNATURE

TODAY LABOR

INITIALS	TRADE	START	FINISH	PAID HOURS	OVERTIME	COMPANY
☐ EMPLOYEE ☐ CONTRUCTOR		AM	PM			
☐ EMPLOYEE ☐ CONTRUCTOR		AM	PM			
☐ EMPLOYEE ☐ CONTRUCTOR		AM	PM			
☐ EMPLOYEE ☐ CONTRUCTOR		AM	PM			
☐ EMPLOYEE ☐ CONTRUCTOR		AM	PM			
☐ EMPLOYEE ☐ CONTRUCTOR		AM	PM			
☐ EMPLOYEE ☐ CONTRUCTOR		AM	PM			
☐ EMPLOYEE ☐ CONTRUCTOR		AM	PM			

EQUIPMENT ON SITE	NO. OF UNITE	WORKING YES / NO

HIRED EQUIPMENT	NO. OF UNITE	EQUIPMENT RENTED	FROM	RATE

NAME: SIGNATURE:

MO TU WE TH FR SA SU
☐   ☐   ☐   ☐   ☐   ☐   ☐

DATE: /   /

PROJECT:

FOREMAN:

WEATHER

F°   C°   AM   PM

| HOURS DUE TO BAD WEATHER | ISSUED AND DELAYS |

NOTE:

COMPLETION DATE	DAYS AHEAD OF SCHEDULE	DAYS BEHIND SCHEDULE

SAFETY AND INCIDENTS

SAFETY ISSUES THAT NEED TO BE ADDRESSED	ACCIDENTS / INCIDENTS / STEPS NEEDED TO RESOLVE

SUMMARY OF THE WORK DONE TODAY

IMPORTANT NOTES

NAME	SIGNATURE

TODAY LABOR

INITIALS	TRADE	START	FINISH	PAID HOURS	OVERTIME	COMPANY
☐ EMPLOYEE ☐ CONTRUCTOR		AM	PM			
☐ EMPLOYEE ☐ CONTRUCTOR		AM	PM			
☐ EMPLOYEE ☐ CONTRUCTOR		AM	PM			
☐ EMPLOYEE ☐ CONTRUCTOR		AM	PM			
☐ EMPLOYEE ☐ CONTRUCTOR		AM	PM			
☐ EMPLOYEE ☐ CONTRUCTOR		AM	PM			
☐ EMPLOYEE ☐ CONTRUCTOR		AM	PM			
☐ EMPLOYEE ☐ CONTRUCTOR		AM	PM			

EQUIPMENT ON SITE	NO. OF UNITE	WORKING YES / NO

HIRED EQUIPMENT	NO. OF UNITE	EQUIPMENT RENTED	FROM	RATE

NAME: _____   SIGNATURE: _____

MO TU WE TH FR SA SU
☐   ☐   ☐   ☐   ☐   ☐   ☐

DATE: ___ / ___ / ___

PROJECT:

FOREMAN:

WEATHER

F°   C°   AM   PM

HOURS DUE TO BAD WEATHER	ISSUED AND DELAYS

NOTE: _____

COMPLETION DATE	DAYS AHEAD OF SCHEDULE	DAYS BEHIND SCHEDULE

SAFETY AND INCIDENTS

SAFETY ISSUES THAT NEED TO BE ADDRESSED	ACCIDENTS / INCIDENTS / STEPS NEEDED TO RESOLVE

SUMMARY OF THE WORK DONE TODAY

IMPORTANT NOTES

NAME	SIGNATURE

TODAY LABOR

INITIALS	TRADE	START	FINISH	PAID HOURS	OVERTIME	COMPANY
☐ EMPLOYEE ☐ CONTRUCTOR		AM	PM			
☐ EMPLOYEE ☐ CONTRUCTOR		AM	PM			
☐ EMPLOYEE ☐ CONTRUCTOR		AM	PM			
☐ EMPLOYEE ☐ CONTRUCTOR		AM	PM			
☐ EMPLOYEE ☐ CONTRUCTOR		AM	PM			
☐ EMPLOYEE ☐ CONTRUCTOR		AM	PM			
☐ EMPLOYEE ☐ CONTRUCTOR		AM	PM			
☐ EMPLOYEE ☐ CONTRUCTOR		AM	PM			

EQUIPMENT ON SITE	NO. OF UNITE	WORKING YES / NO

HIRED EQUIPMENT	NO. OF UNITE	EQUIPMENT RENTED	FROM	RATE

NAME: SIGNATURE:

MO   TU   WE   TH   FR   SA   SU
☐   ☐   ☐   ☐   ☐   ☐   ☐

DATE: __ / __ / __

PROJECT:

FOREMAN:

WEATHER   HOURS DUE TO   ISSUED AND DELAYS
                                           BAD WEATHER

F°____ C°____   AM____ PM____

NOTE: _____

COMPLETION DATE	DAYS AHEAD OF SCHEDULE	DAYS BEHIND SCHEDULE

SAFETY AND INCIDENTS

SAFETY ISSUES THAT NEED TO BE ADDRESSED	ACCIDENTS / INCIDENTS / STEPS NEEDED TO RESOLVE

SUMMARY OF THE WORK DONE TODAY

IMPORTANT NOTES

NAME	SIGNATURE

TODAY LABOR

INITIALS	TRADE	START	FINISH	PAID HOURS	OVERTIME	COMPANY
☐ EMPLOYEE ☐ CONTRUCTOR		AM	PM			
☐ EMPLOYEE ☐ CONTRUCTOR		AM	PM			
☐ EMPLOYEE ☐ CONTRUCTOR		AM	PM			
☐ EMPLOYEE ☐ CONTRUCTOR		AM	PM			
☐ EMPLOYEE ☐ CONTRUCTOR		AM	PM			
☐ EMPLOYEE ☐ CONTRUCTOR		AM	PM			
☐ EMPLOYEE ☐ CONTRUCTOR		AM	PM			
☐ EMPLOYEE ☐ CONTRUCTOR		AM	PM			

EQUIPMENT ON SITE	NO. OF UNITE	WORKING YES / NO

HIRED EQUIPMENT	NO. OF UNITE	EQUIPMENT RENTED	FROM	RATE

NAME: SIGNATURE:

MO   TU   WE   TH   FR   SA   SU
☐   ☐   ☐   ☐   ☐   ☐   ☐

DATE: /   /

PROJECT:

FOREMAN:

WEATHER

F°_____   C°_____   AM_____   PM_____

HOURS DUE TO BAD WEATHER	ISSUED AND DELAYS

NOTE: _____

COMPLETION DATE	DAYS AHEAD OF SCHEDULE	DAYS BEHIND SCHEDULE

SAFETY AND INCIDENTS

SAFETY ISSUES THAT NEED TO BE ADDRESSED	ACCIDENTS / INCIDENTS / STEPS NEEDED TO RESOLVE

SUMMARY OF THE WORK DONE TODAY

IMPORTANT NOTES

NAME	SIGNATURE

TODAY LABOR

INITIALS	TRADE	START	FINISH	PAID HOURS	OVERTIME	COMPANY
☐ EMPLOYEE ☐ CONTRUCTOR		AM	PM			
☐ EMPLOYEE ☐ CONTRUCTOR		AM	PM			
☐ EMPLOYEE ☐ CONTRUCTOR		AM	PM			
☐ EMPLOYEE ☐ CONTRUCTOR		AM	PM			
☐ EMPLOYEE ☐ CONTRUCTOR		AM	PM			
☐ EMPLOYEE ☐ CONTRUCTOR		AM	PM			
☐ EMPLOYEE ☐ CONTRUCTOR		AM	PM			
☐ EMPLOYEE ☐ CONTRUCTOR		AM	PM			

EQUIPMENT ON SITE	NO. OF UNITE	WORKING YES / NO

HIRED EQUIPMENT	NO. OF UNITE	EQUIPMENT RENTED	FROM	RATE

NAME: SIGNATURE:

MO   TU   WE   TH   FR   SA   SU
☐   ☐   ☐   ☐   ☐   ☐   ☐

DATE: __/__/__

PROJECT:

FOREMAN:

WEATHER

F°   C°   AM   PM

HOURS DUE TO BAD WEATHER	ISSUED AND DELAYS

NOTE: _____

COMPLETION DATE	DAYS AHEAD OF SCHEDULE	DAYS BEHIND SCHEDULE

SAFETY AND INCIDENTS

SAFETY ISSUES THAT NEED TO BE ADDRESSED	ACCIDENTS / INCIDENTS / STEPS NEEDED TO RESOLVE

SUMMARY OF THE WORK DONE TODAY

IMPORTANT NOTES

NAME	SIGNATURE

TODAY LABOR

INITIALS	TRADE	START	FINISH	PAID HOURS	OVERTIME	COMPANY
☐ EMPLOYEE ☐ CONTRUCTOR		AM	PM			
☐ EMPLOYEE ☐ CONTRUCTOR		AM	PM			
☐ EMPLOYEE ☐ CONTRUCTOR		AM	PM			
☐ EMPLOYEE ☐ CONTRUCTOR		AM	PM			
☐ EMPLOYEE ☐ CONTRUCTOR		AM	PM			
☐ EMPLOYEE ☐ CONTRUCTOR		AM	PM			
☐ EMPLOYEE ☐ CONTRUCTOR		AM	PM			
☐ EMPLOYEE ☐ CONTRUCTOR		AM	PM			

EQUIPMENT ON SITE	NO. OF UNITE	WORKING YES / NO

HIRED EQUIPMENT	NO. OF UNITE	EQUIPMENT RENTED	FROM	RATE

NAME: 　　　　　　　　　　　SIGNATURE:

MO TU WE TH FR SA SU
☐   ☐   ☐   ☐   ☐   ☐   ☐

DATE: /   /

PROJECT:

FOREMAN:

WEATHER

F°   C°   AM   PM

HOURS DUE TO BAD WEATHER	ISSUED AND DELAYS

NOTE:

COMPLETION DATE	DAYS AHEAD OF SCHEDULE	DAYS BEHIND SCHEDULE

SAFETY AND INCIDENTS

SAFETY ISSUES THAT NEED TO BE ADDRESSED	ACCIDENTS / INCIDENTS / STEPS NEEDED TO RESOLVE

SUMMARY OF THE WORK DONE TODAY

IMPORTANT NOTES

NAME	SIGNATURE

TODAY LABOR

INITIALS	TRADE	START	FINISH	PAID HOURS	OVERTIME	COMPANY
☐ EMPLOYEE ☐ CONTRUCTOR		AM	PM			
☐ EMPLOYEE ☐ CONTRUCTOR		AM	PM			
☐ EMPLOYEE ☐ CONTRUCTOR		AM	PM			
☐ EMPLOYEE ☐ CONTRUCTOR		AM	PM			
☐ EMPLOYEE ☐ CONTRUCTOR		AM	PM			
☐ EMPLOYEE ☐ CONTRUCTOR		AM	PM			
☐ EMPLOYEE ☐ CONTRUCTOR		AM	PM			
☐ EMPLOYEE ☐ CONTRUCTOR		AM	PM			

EQUIPMENT ON SITE	NO. OF UNITE	WORKING YES / NO

HIRED EQUIPMENT	NO. OF UNITE	EQUIPMENT RENTED	FROM	RATE

NAME: SIGNATURE:

MO   TU   WE   TH   FR   SA   SU
☐   ☐   ☐   ☐   ☐   ☐   ☐

DATE: ___/___/___

PROJECT: _____

FOREMAN: _____

WEATHER   ☁   ⛅   ☁   🌨   ☀   🌧   ⛈

F° ___   C° ___   AM ___   PM ___

HOURS DUE TO BAD WEATHER	ISSUED AND DELAYS

NOTE: _____

COMPLETION DATE	DAYS AHEAD OF SCHEDULE	DAYS BEHIND SCHEDULE

SAFETY AND INCIDENTS

SAFETY ISSUES THAT NEED TO BE ADDRESSED	ACCIDENTS / INCIDENTS / STEPS NEEDED TO RESOLVE

SUMMARY OF THE WORK DONE TODAY

IMPORTANT NOTES

NAME	SIGNATURE

TODAY LABOR

INITIALS	TRADE	START	FINISH	PAID HOURS	OVERTIME	COMPANY
☐ EMPLOYEE ☐ CONTRUCTOR		AM	PM			
☐ EMPLOYEE ☐ CONTRUCTOR		AM	PM			
☐ EMPLOYEE ☐ CONTRUCTOR		AM	PM			
☐ EMPLOYEE ☐ CONTRUCTOR		AM	PM			
☐ EMPLOYEE ☐ CONTRUCTOR		AM	PM			
☐ EMPLOYEE ☐ CONTRUCTOR		AM	PM			
☐ EMPLOYEE ☐ CONTRUCTOR		AM	PM			
☐ EMPLOYEE ☐ CONTRUCTOR		AM	PM			

EQUIPMENT ON SITE	NO. OF UNITE	WORKING YES / NO

HIRED EQUIPMENT	NO. OF UNITE	EQUIPMENT RENTED	FROM	RATE

NAME: _____   SIGNATURE: _____

MO   TU   WE   TH   FR   SA   SU
☐   ☐   ☐   ☐   ☐   ☐   ☐   DATE: /   /

PROJECT: FOREMAN:

WEATHER   ☁   ⛅   ☁   🌨   ☀   🌧   ⛈ | HOURS DUE TO BAD WEATHER | ISSUED AND DELAYS |
          F°   C°   AM   PM

NOTE:

COMPLETION DATE	DAYS AHEAD OF SCHEDULE	DAYS BEHIND SCHEDULE

SAFETY AND INCIDENTS

SAFETY ISSUES THAT NEED TO BE ADDRESSED	ACCIDENTS / INCIDENTS / STEPS NEEDED TO RESOLVE

SUMMARY OF THE WORK DONE TODAY

IMPORTANT NOTES

NAME	SIGNATURE

TODAY LABOR

INITIALS	TRADE	START	FINISH	PAID HOURS	OVERTIME	COMPANY
☐ EMPLOYEE ☐ CONTRUCTOR		AM	PM			
☐ EMPLOYEE ☐ CONTRUCTOR		AM	PM			
☐ EMPLOYEE ☐ CONTRUCTOR		AM	PM			
☐ EMPLOYEE ☐ CONTRUCTOR		AM	PM			
☐ EMPLOYEE ☐ CONTRUCTOR		AM	PM			
☐ EMPLOYEE ☐ CONTRUCTOR		AM	PM			
☐ EMPLOYEE ☐ CONTRUCTOR		AM	PM			
☐ EMPLOYEE ☐ CONTRUCTOR		AM	PM			

EQUIPMENT ON SITE	NO. OF UNITE	WORKING YES / NO

HIRED EQUIPMENT	NO. OF UNITE	EQUIPMENT RENTED	FROM	RATE

NAME: SIGNATURE:

MO TU WE TH FR SA SU
☐   ☐   ☐   ☐   ☐   ☐   ☐

DATE: ___ / ___ / ___

PROJECT:

FOREMAN:

WEATHER

F° ___ C° ___   AM ___ PM ___

HOURS DUE TO BAD WEATHER	ISSUED AND DELAYS

NOTE: _____

COMPLETION DATE	DAYS AHEAD OF SCHEDULE	DAYS BEHIND SCHEDULE

SAFETY AND INCIDENTS

SAFETY ISSUES THAT NEED TO BE ADDRESSED	ACCIDENTS / INCIDENTS / STEPS NEEDED TO RESOLVE

SUMMARY OF THE WORK DONE TODAY

IMPORTANT NOTES

NAME	SIGNATURE

TODAY LABOR

INITIALS	TRADE	START	FINISH	PAID HOURS	OVERTIME	COMPANY
☐ EMPLOYEE ☐ CONTRUCTOR		AM	PM			
☐ EMPLOYEE ☐ CONTRUCTOR		AM	PM			
☐ EMPLOYEE ☐ CONTRUCTOR		AM	PM			
☐ EMPLOYEE ☐ CONTRUCTOR		AM	PM			
☐ EMPLOYEE ☐ CONTRUCTOR		AM	PM			
☐ EMPLOYEE ☐ CONTRUCTOR		AM	PM			
☐ EMPLOYEE ☐ CONTRUCTOR		AM	PM			
☐ EMPLOYEE ☐ CONTRUCTOR		AM	PM			

EQUIPMENT ON SITE	NO. OF UNITE	WORKING YES / NO

HIRED EQUIPMENT	NO. OF UNITE	EQUIPMENT RENTED	FROM	RATE

NAME: SIGNATURE:

MO   TU   WE   TH   FR   SA   SU
☐   ☐   ☐   ☐   ☐   ☐   ☐

DATE: /   /

PROJECT:

FOREMAN:

WEATHER

F°   C°   AM   PM

HOURS DUE TO
BAD WEATHER

ISSUED AND DELAYS

NOTE:

COMPLETION DATE	DAYS AHEAD OF SCHEDULE	DAYS BEHIND SCHEDULE

SAFETY AND INCIDENTS

SAFETY ISSUES THAT NEED TO BE ADDRESSED	ACCIDENTS / INCIDENTS / STEPS NEEDED TO RESOLVE

SUMMARY OF THE WORK DONE TODAY

IMPORTANT NOTES

NAME	SIGNATURE

TODAY LABOR

INITIALS	TRADE	START	FINISH	PAID HOURS	OVERTIME	COMPANY
☐ EMPLOYEE ☐ CONTRUCTOR		AM	PM			
☐ EMPLOYEE ☐ CONTRUCTOR		AM	PM			
☐ EMPLOYEE ☐ CONTRUCTOR		AM	PM			
☐ EMPLOYEE ☐ CONTRUCTOR		AM	PM			
☐ EMPLOYEE ☐ CONTRUCTOR		AM	PM			
☐ EMPLOYEE ☐ CONTRUCTOR		AM	PM			
☐ EMPLOYEE ☐ CONTRUCTOR		AM	PM			
☐ EMPLOYEE ☐ CONTRUCTOR		AM	PM			

EQUIPMENT ON SITE	NO. OF UNITE	WORKING YES / NO

HIRED EQUIPMENT	NO. OF UNITE	EQUIPMENT RENTED	FROM	RATE

NAME: SIGNATURE:

MO TU WE TH FR SA SU
☐   ☐   ☐   ☐   ☐   ☐   ☐

DATE: ___/___/___

PROJECT:

FOREMAN:

WEATHER

F°___ C°___   AM___ PM___

HOURS DUE TO BAD WEATHER	ISSUED AND DELAYS

NOTE: _____

COMPLETION DATE	DAYS AHEAD OF SCHEDULE	DAYS BEHIND SCHEDULE

SAFETY AND INCIDENTS

SAFETY ISSUES THAT NEED TO BE ADDRESSED	ACCIDENTS / INCIDENTS / STEPS NEEDED TO RESOLVE

SUMMARY OF THE WORK DONE TODAY

IMPORTANT NOTES

NAME	SIGNATURE

TODAY LABOR

INITIALS	TRADE	START	FINISH	PAID HOURS	OVERTIME	COMPANY
☐ EMPLOYEE ☐ CONTRUCTOR		AM	PM			
☐ EMPLOYEE ☐ CONTRUCTOR		AM	PM			
☐ EMPLOYEE ☐ CONTRUCTOR		AM	PM			
☐ EMPLOYEE ☐ CONTRUCTOR		AM	PM			
☐ EMPLOYEE ☐ CONTRUCTOR		AM	PM			
☐ EMPLOYEE ☐ CONTRUCTOR		AM	PM			
☐ EMPLOYEE ☐ CONTRUCTOR		AM	PM			
☐ EMPLOYEE ☐ CONTRUCTOR		AM	PM			

EQUIPMENT ON SITE	NO. OF UNITE	WORKING YES / NO

HIRED EQUIPMENT	NO. OF UNITE	EQUIPMENT RENTED	FROM	RATE

NAME: _____   SIGNATURE: _____

MO   TU   WE   TH   FR   SA   SU
☐   ☐   ☐   ☐   ☐   ☐   ☐

DATE: /   /

PROJECT:

FOREMAN:

WEATHER

F°   C°   AM   PM

HOURS DUE TO
BAD WEATHER

ISSUED AND DELAYS

NOTE:

COMPLETION DATE	DAYS AHEAD OF SCHEDULE	DAYS BEHIND SCHEDULE

SAFETY AND INCIDENTS

SAFETY ISSUES THAT NEED TO BE ADDRESSED	ACCIDENTS / INCIDENTS / STEPS NEEDED TO RESOLVE

SUMMARY OF THE WORK DONE TODAY

IMPORTANT NOTES

NAME	SIGNATURE

TODAY LABOR

INITIALS	TRADE	START	FINISH	PAID HOURS	OVERTIME	COMPANY
☐ EMPLOYEE ☐ CONTRUCTOR		AM	PM			
☐ EMPLOYEE ☐ CONTRUCTOR		AM	PM			
☐ EMPLOYEE ☐ CONTRUCTOR		AM	PM			
☐ EMPLOYEE ☐ CONTRUCTOR		AM	PM			
☐ EMPLOYEE ☐ CONTRUCTOR		AM	PM			
☐ EMPLOYEE ☐ CONTRUCTOR		AM	PM			
☐ EMPLOYEE ☐ CONTRUCTOR		AM	PM			
☐ EMPLOYEE ☐ CONTRUCTOR		AM	PM			

EQUIPMENT ON SITE	NO. OF UNITE	WORKING YES / NO

HIRED EQUIPMENT	NO. OF UNITE	EQUIPMENT RENTED	FROM	RATE

NAME: SIGNATURE:

MO TU WE TH FR SA SU
☐   ☐   ☐   ☐   ☐   ☐   ☐

DATE: ____ / ____ / ____

PROJECT:

FOREMAN:

WEATHER

F° ____ C° ____   AM ____ PM ____

HOURS DUE TO BAD WEATHER	ISSUED AND DELAYS

NOTE: _____

COMPLETION DATE	DAYS AHEAD OF SCHEDULE	DAYS BEHIND SCHEDULE

SAFETY AND INCIDENTS

SAFETY ISSUES THAT NEED TO BE ADDRESSED	ACCIDENTS / INCIDENTS / STEPS NEEDED TO RESOLVE

SUMMARY OF THE WORK DONE TODAY

IMPORTANT NOTES

NAME	SIGNATURE

TODAY LABOR

INITIALS	TRADE	START	FINISH	PAID HOURS	OVERTIME	COMPANY
☐ EMPLOYEE ☐ CONTRUCTOR		AM	PM			
☐ EMPLOYEE ☐ CONTRUCTOR		AM	PM			
☐ EMPLOYEE ☐ CONTRUCTOR		AM	PM			
☐ EMPLOYEE ☐ CONTRUCTOR		AM	PM			
☐ EMPLOYEE ☐ CONTRUCTOR		AM	PM			
☐ EMPLOYEE ☐ CONTRUCTOR		AM	PM			
☐ EMPLOYEE ☐ CONTRUCTOR		AM	PM			
☐ EMPLOYEE ☐ CONTRUCTOR		AM	PM			

EQUIPMENT ON SITE	NO. OF UNITE	WORKING YES / NO

HIRED EQUIPMENT	NO. OF UNITE	EQUIPMENT RENTED	FROM	RATE

NAME: SIGNATURE:

MO TU WE TH FR SA SU
☐ ☐ ☐ ☐ ☐ ☐ ☐

DATE: ___ / ___ / ___

PROJECT:

FOREMAN:

WEATHER

F° ___ C° ___   AM ___ PM ___

HOURS DUE TO BAD WEATHER	ISSUED AND DELAYS

NOTE: _____

COMPLETION DATE	DAYS AHEAD OF SCHEDULE	DAYS BEHIND SCHEDULE

SAFETY AND INCIDENTS

SAFETY ISSUES THAT NEED TO BE ADDRESSED	ACCIDENTS / INCIDENTS / STEPS NEEDED TO RESOLVE

SUMMARY OF THE WORK DONE TODAY

IMPORTANT NOTES

NAME	SIGNATURE

TODAY LABOR

INITIALS	TRADE	START	FINISH	PAID HOURS	OVERTIME	COMPANY
☐ EMPLOYEE ☐ CONTRUCTOR		AM	PM			
☐ EMPLOYEE ☐ CONTRUCTOR		AM	PM			
☐ EMPLOYEE ☐ CONTRUCTOR		AM	PM			
☐ EMPLOYEE ☐ CONTRUCTOR		AM	PM			
☐ EMPLOYEE ☐ CONTRUCTOR		AM	PM			
☐ EMPLOYEE ☐ CONTRUCTOR		AM	PM			
☐ EMPLOYEE ☐ CONTRUCTOR		AM	PM			
☐ EMPLOYEE ☐ CONTRUCTOR		AM	PM			

EQUIPMENT ON SITE	NO. OF UNITE	WORKING YES / NO

HIRED EQUIPMENT	NO. OF UNITE	EQUIPMENT RENTED	FROM	RATE

NAME:

SIGNATURE:

MO TU WE TH FR SA SU
☐   ☐   ☐   ☐   ☐   ☐   ☐

DATE: ___ / ___ / ___

PROJECT:

FOREMAN:

WEATHER

| HOURS DUE TO BAD WEATHER | ISSUED AND DELAYS |

F° ___ C° ___   ___ AM ___ PM

NOTE: _____

COMPLETION DATE	DAYS AHEAD OF SCHEDULE	DAYS BEHIND SCHEDULE

SAFETY AND INCIDENTS

SAFETY ISSUES THAT NEED TO BE ADDRESSED	ACCIDENTS / INCIDENTS / STEPS NEEDED TO RESOLVE

SUMMARY OF THE WORK DONE TODAY

IMPORTANT NOTES

NAME	SIGNATURE

TODAY LABOR

INITIALS	TRADE	START	FINISH	PAID HOURS	OVERTIME	COMPANY
☐ EMPLOYEE ☐ CONTRUCTOR		AM	PM			
☐ EMPLOYEE ☐ CONTRUCTOR		AM	PM			
☐ EMPLOYEE ☐ CONTRUCTOR		AM	PM			
☐ EMPLOYEE ☐ CONTRUCTOR		AM	PM			
☐ EMPLOYEE ☐ CONTRUCTOR		AM	PM			
☐ EMPLOYEE ☐ CONTRUCTOR		AM	PM			
☐ EMPLOYEE ☐ CONTRUCTOR		AM	PM			
☐ EMPLOYEE ☐ CONTRUCTOR		AM	PM			

EQUIPMENT ON SITE	NO. OF UNITE	WORKING YES / NO	

HIRED EQUIPMENT	NO. OF UNITE	EQUIPMENT RENTED	FROM	RATE

NAME: _____   SIGNATURE: _____

MO TU WE TH FR SA SU
☐   ☐   ☐   ☐   ☐   ☐   ☐

DATE: ___/___/___

PROJECT:

FOREMAN:

WEATHER

F°   C°   AM   PM

HOURS DUE TO BAD WEATHER	ISSUED AND DELAYS

NOTE:

COMPLETION DATE	DAYS AHEAD OF SCHEDULE	DAYS BEHIND SCHEDULE

SAFETY AND INCIDENTS

SAFETY ISSUES THAT NEED TO BE ADDRESSED	ACCIDENTS / INCIDENTS / STEPS NEEDED TO RESOLVE

SUMMARY OF THE WORK DONE TODAY

IMPORTANT NOTES

NAME	SIGNATURE

TODAY LABOR

INITIALS	TRADE	START	FINISH	PAID HOURS	OVERTIME	COMPANY
☐ EMPLOYEE ☐ CONTRUCTOR		AM	PM			
☐ EMPLOYEE ☐ CONTRUCTOR		AM	PM			
☐ EMPLOYEE ☐ CONTRUCTOR		AM	PM			
☐ EMPLOYEE ☐ CONTRUCTOR		AM	PM			
☐ EMPLOYEE ☐ CONTRUCTOR		AM	PM			
☐ EMPLOYEE ☐ CONTRUCTOR		AM	PM			
☐ EMPLOYEE ☐ CONTRUCTOR		AM	PM			
☐ EMPLOYEE ☐ CONTRUCTOR		AM	PM			

EQUIPMENT ON SITE	NO. OF UNITE	WORKING YES / NO

HIRED EQUIPMENT	NO. OF UNITE	EQUIPMENT RENTED	FROM	RATE

NAME: SIGNATURE:

MO   TU   WE   TH   FR   SA   SU
☐   ☐   ☐   ☐   ☐   ☐   ☐

DATE: ___ / ___ / ___

PROJECT:

FOREMAN:

WEATHER

F° ___   C° ___   AM ___   PM ___

HOURS DUE TO BAD WEATHER	ISSUED AND DELAYS

NOTE: _____

COMPLETION DATE	DAYS AHEAD OF SCHEDULE	DAYS BEHIND SCHEDULE

SAFETY AND INCIDENTS

SAFETY ISSUES THAT NEED TO BE ADDRESSED	ACCIDENTS / INCIDENTS / STEPS NEEDED TO RESOLVE

SUMMARY OF THE WORK DONE TODAY

IMPORTANT NOTES

NAME	SIGNATURE

TODAY LABOR

INITIALS	TRADE	START	FINISH	PAID HOURS	OVERTIME	COMPANY
☐ EMPLOYEE ☐ CONTRUCTOR		AM	PM			
☐ EMPLOYEE ☐ CONTRUCTOR		AM	PM			
☐ EMPLOYEE ☐ CONTRUCTOR		AM	PM			
☐ EMPLOYEE ☐ CONTRUCTOR		AM	PM			
☐ EMPLOYEE ☐ CONTRUCTOR		AM	PM			
☐ EMPLOYEE ☐ CONTRUCTOR		AM	PM			
☐ EMPLOYEE ☐ CONTRUCTOR		AM	PM			
☐ EMPLOYEE ☐ CONTRUCTOR		AM	PM			

EQUIPMENT ON SITE	NO. OF UNITE	WORKING YES / NO

HIRED EQUIPMENT	NO. OF UNITE	EQUIPMENT RENTED	FROM	RATE

NAME: _____   SIGNATURE: _____

MO   TU   WE   TH   FR   SA   SU
☐   ☐   ☐   ☐   ☐   ☐   ☐

DATE: /   /

PROJECT:

FOREMAN:

WEATHER

F°____ C°____   AM____ PM____

HOURS DUE TO BAD WEATHER	ISSUED AND DELAYS

NOTE: _____

COMPLETION DATE	DAYS AHEAD OF SCHEDULE	DAYS BEHIND SCHEDULE

SAFETY AND INCIDENTS

SAFETY ISSUES THAT NEED TO BE ADDRESSED	ACCIDENTS / INCIDENTS / STEPS NEEDED TO RESOLVE

SUMMARY OF THE WORK DONE TODAY

IMPORTANT NOTES

NAME	SIGNATURE

TODAY LABOR

INITIALS	TRADE	START	FINISH	PAID HOURS	OVERTIME	COMPANY
☐ EMPLOYEE ☐ CONTRUCTOR		AM	PM			
☐ EMPLOYEE ☐ CONTRUCTOR		AM	PM			
☐ EMPLOYEE ☐ CONTRUCTOR		AM	PM			
☐ EMPLOYEE ☐ CONTRUCTOR		AM	PM			
☐ EMPLOYEE ☐ CONTRUCTOR		AM	PM			
☐ EMPLOYEE ☐ CONTRUCTOR		AM	PM			
☐ EMPLOYEE ☐ CONTRUCTOR		AM	PM			
☐ EMPLOYEE ☐ CONTRUCTOR		AM	PM			

EQUIPMENT ON SITE	NO. OF UNITE	WORKING YES / NO

HIRED EQUIPMENT	NO. OF UNITE	EQUIPMENT RENTED	FROM	RATE

NAME: SIGNATURE:

MO   TU   WE   TH   FR   SA   SU
☐   ☐   ☐   ☐   ☐   ☐   ☐

DATE: ____ / ____ / ____

PROJECT:

FOREMAN:

WEATHER

F° ____   C° ____   ____ AM   ____ PM

HOURS DUE TO BAD WEATHER	ISSUED AND DELAYS

NOTE: _____

COMPLETION DATE	DAYS AHEAD OF SCHEDULE	DAYS BEHIND SCHEDULE

SAFETY AND INCIDENTS

SAFETY ISSUES THAT NEED TO BE ADDRESSED	ACCIDENTS / INCIDENTS / STEPS NEEDED TO RESOLVE

SUMMARY OF THE WORK DONE TODAY

IMPORTANT NOTES

NAME	SIGNATURE

TODAY LABOR

INITIALS	TRADE	START	FINISH	PAID HOURS	OVERTIME	COMPANY
☐ EMPLOYEE ☐ CONTRUCTOR		AM	PM			
☐ EMPLOYEE ☐ CONTRUCTOR		AM	PM			
☐ EMPLOYEE ☐ CONTRUCTOR		AM	PM			
☐ EMPLOYEE ☐ CONTRUCTOR		AM	PM			
☐ EMPLOYEE ☐ CONTRUCTOR		AM	PM			
☐ EMPLOYEE ☐ CONTRUCTOR		AM	PM			
☐ EMPLOYEE ☐ CONTRUCTOR		AM	PM			
☐ EMPLOYEE ☐ CONTRUCTOR		AM	PM			

EQUIPMENT ON SITE	NO. OF UNITE	WORKING YES / NO

HIRED EQUIPMENT	NO. OF UNITE	EQUIPMENT RENTED	FROM	RATE

NAME: SIGNATURE:

MO TU WE TH FR SA SU
☐   ☐   ☐   ☐   ☐   ☐   ☐

DATE: /   /

PROJECT:

FOREMAN:

WEATHER

F°____   C°____   AM____   PM____

HOURS DUE TO BAD WEATHER	ISSUED AND DELAYS

NOTE: _____

COMPLETION DATE	DAYS AHEAD OF SCHEDULE	DAYS BEHIND SCHEDULE

SAFETY AND INCIDENTS

SAFETY ISSUES THAT NEED TO BE ADDRESSED	ACCIDENTS / INCIDENTS / STEPS NEEDED TO RESOLVE

SUMMARY OF THE WORK DONE TODAY

IMPORTANT NOTES

NAME	SIGNATURE

TODAY LABOR

INITIALS	TRADE	START	FINISH	PAID HOURS	OVERTIME	COMPANY
☐ EMPLOYEE ☐ CONTRUCTOR		AM	PM			
☐ EMPLOYEE ☐ CONTRUCTOR		AM	PM			
☐ EMPLOYEE ☐ CONTRUCTOR		AM	PM			
☐ EMPLOYEE ☐ CONTRUCTOR		AM	PM			
☐ EMPLOYEE ☐ CONTRUCTOR		AM	PM			
☐ EMPLOYEE ☐ CONTRUCTOR		AM	PM			
☐ EMPLOYEE ☐ CONTRUCTOR		AM	PM			
☐ EMPLOYEE ☐ CONTRUCTOR		AM	PM			

EQUIPMENT ON SITE	NO. OF UNITE	WORKING YES / NO

HIRED EQUIPMENT	NO. OF UNITE	EQUIPMENT RENTED	FROM	RATE

NAME: SIGNATURE:

MO TU WE TH FR SA SU
☐   ☐   ☐   ☐   ☐   ☐   ☐

DATE: /   /

PROJECT:

FOREMAN:

WEATHER

F°   C°   AM   PM

| HOURS DUE TO BAD WEATHER | ISSUED AND DELAYS |

NOTE: _____

COMPLETION DATE	DAYS AHEAD OF SCHEDULE	DAYS BEHIND SCHEDULE

SAFETY AND INCIDENTS

SAFETY ISSUES THAT NEED TO BE ADDRESSED	ACCIDENTS / INCIDENTS / STEPS NEEDED TO RESOLVE

SUMMARY OF THE WORK DONE TODAY

IMPORTANT NOTES

NAME	SIGNATURE

TODAY LABOR

INITIALS	TRADE	START	FINISH	PAID HOURS	OVERTIME	COMPANY
☐ EMPLOYEE ☐ CONTRUCTOR		AM	PM			
☐ EMPLOYEE ☐ CONTRUCTOR		AM	PM			
☐ EMPLOYEE ☐ CONTRUCTOR		AM	PM			
☐ EMPLOYEE ☐ CONTRUCTOR		AM	PM			
☐ EMPLOYEE ☐ CONTRUCTOR		AM	PM			
☐ EMPLOYEE ☐ CONTRUCTOR		AM	PM			
☐ EMPLOYEE ☐ CONTRUCTOR		AM	PM			
☐ EMPLOYEE ☐ CONTRUCTOR		AM	PM			

EQUIPMENT ON SITE	NO. OF UNITE	WORKING YES / NO

HIRED EQUIPMENT	NO. OF UNITE	EQUIPMENT RENTED	FROM	RATE

NAME: SIGNATURE:

MO TU WE TH FR SA SU
☐ ☐ ☐ ☐ ☐ ☐ ☐

DATE: /   /

PROJECT:

FOREMAN:

WEATHER
F°   C°   AM   PM

HOURS DUE TO
BAD WEATHER

ISSUED AND DELAYS

NOTE:

COMPLETION DATE	DAYS AHEAD OF SCHEDULE	DAYS BEHIND SCHEDULE

SAFETY AND INCIDENTS

SAFETY ISSUES THAT NEED TO BE ADDRESSED	ACCIDENTS / INCIDENTS / STEPS NEEDED TO RESOLVE

SUMMARY OF THE WORK DONE TODAY

IMPORTANT NOTES

NAME	SIGNATURE

TODAY LABOR

INITIALS	TRADE	START	FINISH	PAID HOURS	OVERTIME	COMPANY
☐ EMPLOYEE ☐ CONTRUCTOR		AM	PM			
☐ EMPLOYEE ☐ CONTRUCTOR		AM	PM			
☐ EMPLOYEE ☐ CONTRUCTOR		AM	PM			
☐ EMPLOYEE ☐ CONTRUCTOR		AM	PM			
☐ EMPLOYEE ☐ CONTRUCTOR		AM	PM			
☐ EMPLOYEE ☐ CONTRUCTOR		AM	PM			
☐ EMPLOYEE ☐ CONTRUCTOR		AM	PM			
☐ EMPLOYEE ☐ CONTRUCTOR		AM	PM			

EQUIPMENT ON SITE	NO. OF UNITE	WORKING YES / NO

HIRED EQUIPMENT	NO. OF UNITE	EQUIPMENT RENTED	FROM	RATE

NAME: SIGNATURE:

MO　TU　WE　TH　FR　SA　SU
☐　　☐　　☐　　☐　　☐　　☐　　☐

DATE: ___ / ___ / ___

PROJECT:

FOREMAN:

WEATHER

F°___　C°___　___ AM ___ PM

HOURS DUE TO BAD WEATHER	ISSUED AND DELAYS

NOTE:

COMPLETION DATE	DAYS AHEAD OF SCHEDULE	DAYS BEHIND SCHEDULE

SAFETY AND INCIDENTS

SAFETY ISSUES THAT NEED TO BE ADDRESSED	ACCIDENTS / INCIDENTS / STEPS NEEDED TO RESOLVE

SUMMARY OF THE WORK DONE TODAY

IMPORTANT NOTES

NAME	SIGNATURE

TODAY LABOR

INITIALS	TRADE	START	FINISH	PAID HOURS	OVERTIME	COMPANY
☐ EMPLOYEE ☐ CONTRUCTOR		AM	PM			
☐ EMPLOYEE ☐ CONTRUCTOR		AM	PM			
☐ EMPLOYEE ☐ CONTRUCTOR		AM	PM			
☐ EMPLOYEE ☐ CONTRUCTOR		AM	PM			
☐ EMPLOYEE ☐ CONTRUCTOR		AM	PM			
☐ EMPLOYEE ☐ CONTRUCTOR		AM	PM			
☐ EMPLOYEE ☐ CONTRUCTOR		AM	PM			
☐ EMPLOYEE ☐ CONTRUCTOR		AM	PM			

EQUIPMENT ON SITE	NO. OF UNITE	WORKING YES / NO

HIRED EQUIPMENT	NO. OF UNITE	EQUIPMENT RENTED	FROM	RATE

NAME: SIGNATURE:

MO TU WE TH FR SA SU
☐   ☐   ☐   ☐   ☐   ☐   ☐

DATE: / /

PROJECT:

FOREMAN:

WEATHER

F°   C°   AM   PM

HOURS DUE TO BAD WEATHER	ISSUED AND DELAYS

NOTE:

COMPLETION DATE	DAYS AHEAD OF SCHEDULE	DAYS BEHIND SCHEDULE

SAFETY AND INCIDENTS

SAFETY ISSUES THAT NEED TO BE ADDRESSED	ACCIDENTS / INCIDENTS / STEPS NEEDED TO RESOLVE

SUMMARY OF THE WORK DONE TODAY

IMPORTANT NOTES

NAME	SIGNATURE

TODAY LABOR

INITIALS	TRADE	START	FINISH	PAID HOURS	OVERTIME	COMPANY
☐ EMPLOYEE ☐ CONTRUCTOR		AM	PM			
☐ EMPLOYEE ☐ CONTRUCTOR		AM	PM			
☐ EMPLOYEE ☐ CONTRUCTOR		AM	PM			
☐ EMPLOYEE ☐ CONTRUCTOR		AM	PM			
☐ EMPLOYEE ☐ CONTRUCTOR		AM	PM			
☐ EMPLOYEE ☐ CONTRUCTOR		AM	PM			
☐ EMPLOYEE ☐ CONTRUCTOR		AM	PM			
☐ EMPLOYEE ☐ CONTRUCTOR		AM	PM			

EQUIPMENT ON SITE	NO. OF UNITE	WORKING YES / NO

HIRED EQUIPMENT	NO. OF UNITE	EQUIPMENT RENTED	FROM	RATE

NAME: SIGNATURE:

MO TU WE TH FR SA SU
☐   ☐   ☐   ☐   ☐   ☐   ☐

DATE: ___ / ___ / ___

PROJECT:

FOREMAN:

WEATHER

F°___ C°___ ___ AM ___ PM

HOURS DUE TO BAD WEATHER

ISSUED AND DELAYS

NOTE:

COMPLETION DATE	DAYS AHEAD OF SCHEDULE	DAYS BEHIND SCHEDULE

SAFETY AND INCIDENTS

SAFETY ISSUES THAT NEED TO BE ADDRESSED	ACCIDENTS / INCIDENTS / STEPS NEEDED TO RESOLVE

SUMMARY OF THE WORK DONE TODAY

IMPORTANT NOTES

NAME	SIGNATURE

TODAY LABOR

INITIALS	TRADE	START	FINISH	PAID HOURS	OVERTIME	COMPANY
☐ EMPLOYEE ☐ CONTRUCTOR		AM	PM			
☐ EMPLOYEE ☐ CONTRUCTOR		AM	PM			
☐ EMPLOYEE ☐ CONTRUCTOR		AM	PM			
☐ EMPLOYEE ☐ CONTRUCTOR		AM	PM			
☐ EMPLOYEE ☐ CONTRUCTOR		AM	PM			
☐ EMPLOYEE ☐ CONTRUCTOR		AM	PM			
☐ EMPLOYEE ☐ CONTRUCTOR		AM	PM			
☐ EMPLOYEE ☐ CONTRUCTOR		AM	PM			

EQUIPMENT ON SITE	NO. OF UNITE	WORKING YES / NO	

HIRED EQUIPMENT	NO. OF UNITE	EQUIPMENT RENTED	FROM	RATE

NAME: _____   SIGNATURE: _____

MO   TU   WE   TH   FR   SA   SU
☐   ☐   ☐   ☐   ☐   ☐   ☐

DATE: /   /

PROJECT:

FOREMAN:

WEATHER

F°_____ C°_____   AM_____ PM_____

HOURS DUE TO BAD WEATHER	ISSUED AND DELAYS

NOTE: _____

COMPLETION DATE	DAYS AHEAD OF SCHEDULE	DAYS BEHIND SCHEDULE

SAFETY AND INCIDENTS

SAFETY ISSUES THAT NEED TO BE ADDRESSED	ACCIDENTS / INCIDENTS / STEPS NEEDED TO RESOLVE

SUMMARY OF THE WORK DONE TODAY

IMPORTANT NOTES

NAME	SIGNATURE

TODAY LABOR

INITIALS	TRADE	START	FINISH	PAID HOURS	OVERTIME	COMPANY
☐ EMPLOYEE ☐ CONTRUCTOR		AM	PM			
☐ EMPLOYEE ☐ CONTRUCTOR		AM	PM			
☐ EMPLOYEE ☐ CONTRUCTOR		AM	PM			
☐ EMPLOYEE ☐ CONTRUCTOR		AM	PM			
☐ EMPLOYEE ☐ CONTRUCTOR		AM	PM			
☐ EMPLOYEE ☐ CONTRUCTOR		AM	PM			
☐ EMPLOYEE ☐ CONTRUCTOR		AM	PM			
☐ EMPLOYEE ☐ CONTRUCTOR		AM	PM			

EQUIPMENT ON SITE	NO. OF UNITE	WORKING YES / NO

HIRED EQUIPMENT	NO. OF UNITE	EQUIPMENT RENTED	FROM	RATE

NAME: SIGNATURE:

MO TU WE TH FR SA SU
☐   ☐   ☐   ☐   ☐   ☐   ☐

DATE: /   /

PROJECT:

FOREMAN:

WEATHER

F°____ C°____   ____ AM ____ PM

HOURS DUE TO BAD WEATHER	ISSUED AND DELAYS

NOTE: _____

COMPLETION DATE	DAYS AHEAD OF SCHEDULE	DAYS BEHIND SCHEDULE

SAFETY AND INCIDENTS

SAFETY ISSUES THAT NEED TO BE ADDRESSED	ACCIDENTS / INCIDENTS / STEPS NEEDED TO RESOLVE

SUMMARY OF THE WORK DONE TODAY

IMPORTANT NOTES

NAME	SIGNATURE

TODAY LABOR

INITIALS	TRADE	START	FINISH	PAID HOURS	OVERTIME	COMPANY
☐ EMPLOYEE ☐ CONTRUCTOR		AM	PM			
☐ EMPLOYEE ☐ CONTRUCTOR		AM	PM			
☐ EMPLOYEE ☐ CONTRUCTOR		AM	PM			
☐ EMPLOYEE ☐ CONTRUCTOR		AM	PM			
☐ EMPLOYEE ☐ CONTRUCTOR		AM	PM			
☐ EMPLOYEE ☐ CONTRUCTOR		AM	PM			
☐ EMPLOYEE ☐ CONTRUCTOR		AM	PM			
☐ EMPLOYEE ☐ CONTRUCTOR		AM	PM			

EQUIPMENT ON SITE	NO. OF UNITE	WORKING YES / NO

HIRED EQUIPMENT	NO. OF UNITE	EQUIPMENT RENTED	FROM	RATE

NAME: SIGNATURE:

IMPORTANT TELEPHONE NUMBER

▷ NAME .... PHONE ....
 EMAIL ....

▷ NAME .... PHONE ....
 EMAIL ....

▷ NAME .... PHONE ....
 EMAIL ....

▷ NAME .... PHONE ....
 EMAIL ....

▷ NAME .... PHONE ....
 EMAIL ....

▷ NAME .... PHONE ....
 EMAIL ....

▷ NAME .... PHONE ....
 EMAIL ....

▷ NAME .... PHONE ....
 EMAIL ....

▷ NAME .... PHONE ....
 EMAIL ....

▷ NAME .... PHONE ....
 EMAIL ....

▷ NAME .... PHONE ....
 EMAIL ....

▷ NAME .... PHONE ....
 EMAIL ....

▷ NAME .... PHONE ....
 EMAIL ....

▷ NAME .... PHONE ....
 EMAIL ....